Urban Traffic Safety: Research Proposals for Tashkent

Xamroqulov R.M.

Tashkent-2026

Urban Traffic Safety: Research Proposals for Tashkent

By: Xamroqulov R.M.

Edition: June '2026

Publisher:

Taemeer Publications LLC (Michigan, USA / Hyderabad, India)

ISBN 978-93-6908-023-6

9 789369 080236

Book	:	**Urban Traffic Safety: Research Proposals for Tashkent**
Editor	:	Xamroqulov R.M.
Publisher	:	Taemeer Publications
Year	:	'2026
Pages	:	104
Title Design	:	*Taemeer Web Design*

Distributed globally through IngramSpark
Printed on demand by Ingram Content Group through regional printing facilities depending on order destination.

SIGN IN

The use of technical means for organizing and regulating traffic is an important topic in today's rapidly interconnected world. With rapid development, population growth, and the growing demand for efficient transport systems, the need for efficient vehicle management has never been more important, and technological advancements have revolutionized the way we approach and regulate many aspects of our daily lives. One of the areas where technical means play a decisive role is the organization and regulation of movement. Technological solutions ranging from traffic management systems to logistics and transport management have significantly increased the efficiency, safety, and overall efficiency of traffic management. In accordance with the Strategy for the Development of the Public Safety System in the

Republic of Uzbekistan for 2022-2025, and for the purpose of guaranteeing the protection of human life and health from any road accidents under the conditions of New Uzbekistan:

We define the following as modern directions for ensuring road safety in our country:

improving and improving the quality of road infrastructure, creating reliable conditions for the safe movement of road users based on the priority of pedestrians - public transport - bicycle transport - road transport;

bringing the educational process to a qualitatively new level by introducing innovative pedagogical technologies into the system of training, retraining, and advanced training of drivers;

increasing the culture of compliance with traffic rules among drivers and pedestrians, ensuring the inevitability of punishment for any violation;

Establishing the introduction of the basics of traffic rules from childhood, introducing this practice in preschool educational organizations and general education schools;

full digitalization of traffic management, the introduction of new management and control systems, and the introduction of advanced information and communication technologies.

In order to ensure strict compliance with traffic rules and prevent their repeated violations, a system for calculating fines for drivers for traffic violations was introduced on December 1, 2022, which provides for the following procedure:

for each traffic violation, separate penalty points are assigned for 12 months;

in cases where the accrued penalty points exceed the established amount, the driver shall be deprived of the right to drive the vehicle for a certain period in accordance with the established procedure;

CHAPTER 1. APPLICATION OF TECHNICAL MEANS FOR MOVEMENT ORGANIZATION AND REGULATION

1.1 Requirements for the installation of road signs

The placement of road signs on highways and city streets must comply with the requirements of GOST 23457-86, and their dimensions must comply with the requirements of GOST 10807-78. Road signs No. 1 must be placed in the following order: sign (informative sign); privilege; service plates; warning; command; prohibition [9].

The total number of installed road signs on the road should be as small as possible. Each placed warning or prohibiting road sign must be justified and understood as a compulsory measure taken due to the deterioration of road conditions. Therefore, in most cases, warning and prohibitory signs must be temporarily installed and removed after the condition of the road under consideration has improved. Furthermore, seasonal road signs must be removed quickly over time.

There should be no strong obstacles in front of or near the road sign that prevent people from seeing it, and one should try not to be frequently blocked by other temporary objects [6].

Warning signs.

Warning signs are used to inform road users about the location and characteristics of hazardous areas requiring action depending on the situation.

There are 31 warning signs, the majority of which feature a triangle, a red border, a white background, and a black image. In addition, it is available in the form of rectangles and segments. Warning signs are installed in populated areas

50-100 meters before the danger zone and 150-300 meters outside of populated areas.

If the slope and height are consecutive, signs may be placed immediately before the slope and height. The second sign is installed at least 50 meters before the start of the dangerous section of the road.

If short-term road works have been carried out on the roadway, the required sign is installed at a distance of 10–15 meters from the work site.

Road signs, traffic lights, barriers, and direction finding devices must be positioned to ensure optimal visibility for road users around the clock, ease of use and maintenance, and to eliminate the possibility of intentional damage. At the same time, they must not be blocked from road users by any obstacles (green spaces, external lighting poles, etc.).

When placing road signs and traffic lights, it is necessary to ensure that the information they convey is directed only at the road users intended for them [11].

On road sections where road signs determining the traffic regime are difficult to distinguish (snow, mud, etc.) or cannot be restored, road signs corresponding to their meaning must be installed.

The placement of advertisements, posters, banners, and other devices on roads and outside of populated areas is permitted if it does not interfere with the perception of technical means of traffic regulation by participants.

The placement of road signs, traffic lights and their supports, advertising, posters, banners, and other devices not related to traffic management is not permitted.

Technical means of traffic control put into operation for temporary reasons (repair of roads, seasonal traffic regime, etc.) must be dismantled immediately after eliminating the specified reasons. In this case, signs and traffic lights may be covered with a cover.

Technical means of traffic organization may be used in cases not provided for by this standard if the necessity of their use is justified by specific traffic conditions.

The use of technical means of traffic control for experimental purposes not regulated by current state standards is permitted. In this case, if necessary, road users must be provided with information explaining the purpose of this technical device.

Requirements for road signs include:

The numbers, names, and images of road signs are provided in Appendix No. 1.

The sign is located near or above a road, bike path, or pedestrian path.

On non-stationary illuminated road sections and on stationary illuminated road sections, signs with reflective surfaces—internal and external illumination—must be used.

On road sections with stationary lighting, the use of signs with a reflective surface is permitted if the signs are visible from a distance of at least 100 m [12].

No more than three signs may be installed on a single section of the road, with the exception of duplicate signs and additional information signs (posters).

Signs installed sequentially on the road, with the exception of signs installed at intersections, must be located at a distance of at least 50 m from settlements and at least 25 m from each other in settlements.

Road signs must be installed on the right side of the road and behind the shoulder, unless otherwise provided by this standard. For one-way traffic on roads with two or more sections, if, taking into account the nature of the information contained therein and local conditions, drivers did not have time to become aware of the movement of large vehicles, signs must be duplicated on the right-hand side of the road. Signs in duplicate must be installed on the dividing line, on the left side of the road if there is no dividing line, on the upper part of the road if there are no more than two lanes for traffic in the opposite direction, or on the upper part of the road if a visibility sign is installed. the left side of the road is not provided for, but there are three or more lanes for traffic in the opposite direction.

The distance from the edge of the road, if there is a shoulder, should be from 0.5 to 2 m from the edge of the road to the nearest edge of the sign installed at the edge of the road. 5.20.1, 5.21.1 - 5.27, 5.31 - from 0.5 to 5.0 m to the edge of information signs.

In narrow road conditions (near rocks, boulders, parapets, etc.), it is permitted to install signs along the edges of the road. In this case, the distance between the edge of the road and the edge of the nearest sign must be at least 1 m, and the installation height must be at least 2 m.

In the absence of a fence, signs must be installed on safe supports (according to GOST 25458 or GOST 25459) at the edge of the road, as well as on the dividing line. The upper edge of the foundation supporting the sign must be identical to the edge, median, or berma surface.

Signs can be installed on mobile supports in places where they work on the road and when traffic changes on the road [5].

The distance from the lower edge of the sign (except for warning signs and slabs 1.4.1-1.4.6) to the road surface (installation height), except in cases specifically provided for by this standard. :

- from 1.5 to 2.2 m - when installed on the roadside outside of populated areas, from 2.0 to 4.0 m - in populated areas;

- at least 0.6 m - when installed on transport islands and on the road;

- from 5.0 to 6.0 m - when located in the upper part of the road; If the distance from the road surface to the lower part of the structure is less than 5 m, signs must not extend beyond the lower edge of artificial structures when placing them between them.

The height of roadside markings is determined from the surface of the road surface at the edge of the road.

If the signs are located beneath each other, the installation height is indicated by a lower line.

Signs must be as high as possible on the same road.

The distance between adjacent signs installed on the same support and distributing their impact in the same direction must be from 50 to 200 mm, with the exception of signs made on the same housing.

Signs must be installed at a distance of no less than 1 m from high-voltage network wires. It is prohibited to mark the guy wires within the safety zone of high-voltage lines.

The standard dimensions of the signs and the height of the capital letters for individual information signs must be selected in accordance with GOST 10807 depending on the conditions of use. Depending on traffic conditions, signs of the nearest standard size must be used.

Warning signs

Warning signs serve to inform road users about the nature of the danger and the location of a dangerous road section requiring action appropriate to the situation.

Outside settlements, in addition to signs 1.3.1 - 1.4.6, 1.31.1 - 1.31.3, warning signs must be installed depending on the speed of movement, visibility conditions, and placement capabilities. from 150 to 300 m, and in populated areas from 50 to 100 m from the beginning of the hazardous zone.

If necessary (if the beginning of a dangerous section is hidden from the driver by a sharp turn, a longitudinal profile curve, a building, etc.), it is permitted to install warning signs at a different distance specified in this case.

It is permitted to install a second warning sign with sign 7.1.1 if there is an intersection of the warning sign and the start of the dangerous section outside this intersection. If the distance between the start of the hazardous section and the intersection is less than 20 m, it is permitted to install the corresponding sign with slabs 7.1.3 or 7.1.4 at the intersection at a distance of up to 50 m.

Signs 1.1, 1.2, 1.9, 1.10, 1.21, and 1.23 must be repeated outside settlements, while signs 1.21, and 1.23 must be repeated within settlements. The second sign must be installed at a distance of at least 50 m from the start of the hazardous section.

Signs 1.1 for railway crossings with barriers and 1.2 for railway crossings without barriers are installed in front of all railway crossings equipped with barriers or without them, respectively.

Signs 1.1 and 1.2 are used for traffic in both directions on roads with three or more lanes, as well as one for traffic in both directions if the crossing's visibility outside of populated areas is less than 300 m or must be repeated on two-lane roads, and less than 100 m in populated areas [3].

If the road crosses crossings with a distance of less than 50 m between them, signs 1.1 or 1.2 must be installed only before the first crossing, and in other cases, before each crossing.

If the tracks are parallel to the railway tracks for a distance of 20 to 50 m, signs 1.1 or 1.2 with plates 7.1.1 should be additionally installed on the tracks crossing them.

If the distance between lanes is less than 20 m, signs 1.1 or 1.2 with slabs 7.1.3 or 7.1.4 must be installed at a distance of 50 m from the intersection with the road leading to the intersection.

1.3.1 All unobstructed railway crossings along the railway track must be installed before one, two, or more or more of them are installed, respectively. 1.3.2 Single-track railway and multi-track railway signs. If a traffic light is present at an intersection, signs 1.3.1 and 1.3.2 must be installed on the same pole as the traffic light, or in the absence of a traffic light, at a distance of at least 20 m from the nearest railway track.

1.4.1-1.4.6 Signs for approaching a railway crossing must be installed on highways of republican, republican, and regional significance outside of populated areas in front of each crossing, as well as on other roads provided there is visibility from the crossing. Less than 300 m.

Signs 1.4.1-1.4.3 must be installed on the right side of the road, and signs 1.4.4-1.4.6 on the left. Signs 1.4.1 and 1.4.4 must be installed along the direction of travel of the first (main and backup) signs 1.1 or 1.2, and signs 1.4.3 and 1.4.6 along the direction of travel of the second signs 1.1 or 1.2 and 1.4. 2 and 1.4.5 - independently, at equal distances between the first and second signs 1.1 or 1.2.

If the road is parallel to tram tracks at a distance of less than 20 m, a sign 1.5 with slabs 7.1.3 or 7.1.4 must be installed 50 m from the intersection before the intersection with the road leading to the tram tracks.

The movement of enterprises located in the designated territory, as well as other vehicles belonging to citizens or citizens residing in the designated territory, is permitted. Such vehicles must enter and exit the designated area at the intersection closest to the attraction.

The total number of installed road signs on the road should be as small as possible. Each placed warning or prohibiting road sign must be justified and understood as a compulsory measure taken due to the deterioration of road conditions. Therefore, in most cases, warning and prohibitory signs must be temporarily installed and removed after certain road conditions have improved. Seasonal road signs should also be removed quickly over time.

Road signs must be installed so that their visibility during daylight hours is at least 150 m, while the distance at which a driver can see a sign installed in a specific area must be equal to or greater than twice the speed in that area. At the same time, on newly constructed roads, the speed will be 70% of the design speed, while on existing roads, it will not exceed 85% of the vehicle speed.

In some cases, when it is necessary to determine the distance between the installed sign and the hazardous zone, it can be determined using the following diagrams and formulas [10].

Depending on the speed of fast-moving vehicles in the area, it changes according to the formula given in the table below.

Table 1.1.

Average speed of fast cars						
V_{yp}	km/h	120	100	80	60	40
l	m	254	200	147	99	56

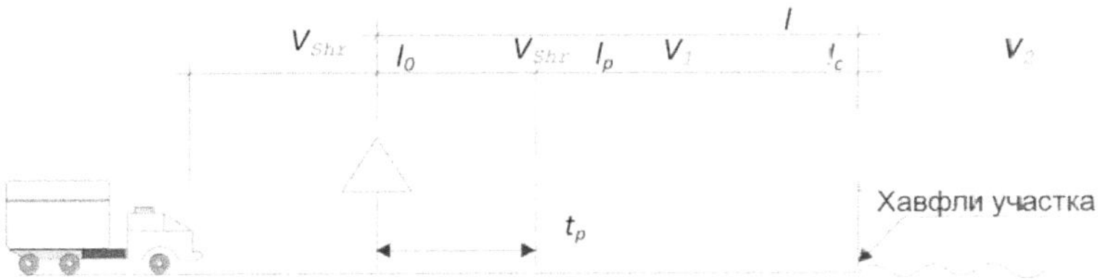

Figure 1.1. A road sign placed before a dangerous section
distance determination scheme

Traffic police officers must regularly clear road signs and cut down trees that obstruct visibility. Single-row road signs must be installed at a distance of at least 50 m on non-residential road sections and 25 m in populated areas. This does not include road signs at intersections.

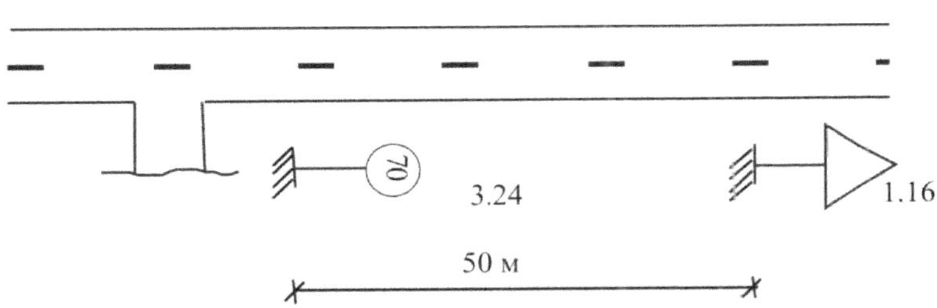

1.2-Fig. Planning road signs in non-residential areas

When placing road signs in cross-sections, the following must be taken into account. The installation of more than three road signs on a single road section is not permitted. Additional information and returned characters are not taken into account. In such cases, signs should be installed either horizontally (which is more convenient) or vertically.

Road signs are installed on the right side of the roadway and beyond the shoulder (except in exceptional cases).

It is permitted to install road signs on the road (for example, on streets around the Pakhtakor Stadium) during repair or repair work or in an operational (operational) state, or temporary signs prohibiting the movement of these vehicles or their movement at all).

In settlements, road signs may be placed on the edge of the road (Fig. 1.4a), above the edge of the road, or in the upper part of the road (Fig.

1.4). In cases where traffic islands are located on the dividing line, the placement of the traffic sign is shown in Fig. 84g.

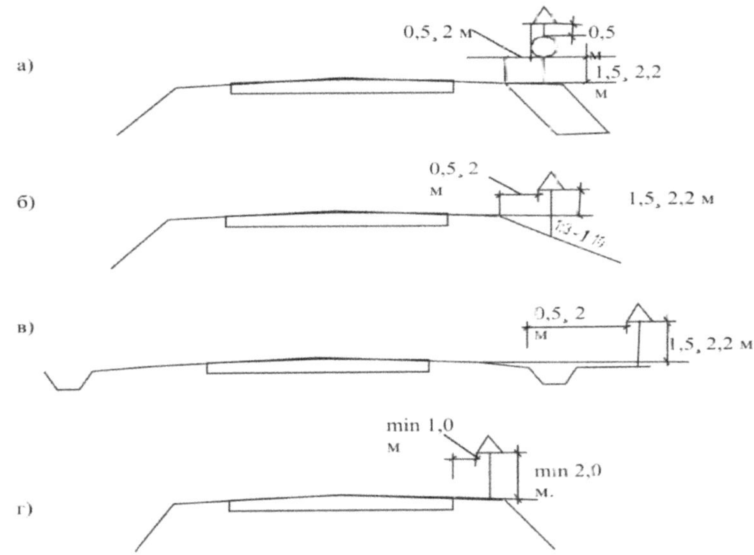

Figure 1.3. Cross the road signs in uninhabited areas. placement in a cross-section.

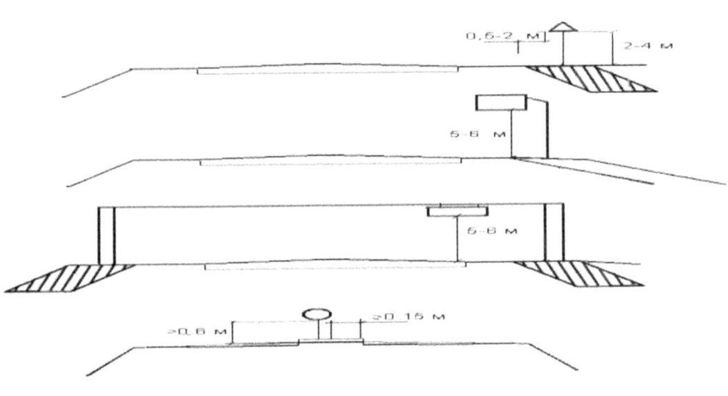

Figure 1.4. Cross the road signs in residential areas placement in the cross-section.

Dimensions of the first type of additional information signs = 600 mm; = 900 mm. The full dimensions of all marks are given in GOST 10807-78.

The support of the sign is buried in the ground to a depth of at least 1.2 m; if this depth is less than 1.2 m, concrete is laid in a circle around the buried part of the support.

Warning road signs inform the driver about dangerous areas in front of the road.

Places of use for road signs by type.

Table 1.2.

Character size types	Place of use of a road sign	
	In non-residential areas	Populated in points
I-small	On a single lane road	On local roads
II-normal	On two-and three-lane roads	Highway on the roads
III-major	On four-lane and multi-lane roads and highways	Fast-moving on the roads

IV-very large	In cases where the effectiveness of use is justified in areas where repair work is being carried out on highways and on dangerous sections of other roads

Depending on the type of dimensions, triangular (equilateral) road signs are adopted as follows.

Below we provide some guidance on how to use traffic warning signs in different road conditions.

Road sign Equivalent road intersection 1.6 shall be installed at intersections where the visibility distance is less than 150 m in non-residential areas and less than 50 m in populated areas.

Before the organization of each bypass road and the installation of sign 4.3, the first bypass road sign 1.7 should be installed in settlements with an intersection visibility of less than 50 m, as well as at unlit intersections.

Signs 1.11.1 and 1.11.2 for dangerous turns must be installed before the planned turn if the safety factor is 0.6, as well as if the visibility range of vehicles approaching the turn is lower than the value specified in this article.

Signs for hazardous turns 1.12.1 and 1.12.2 must be placed before two or more consecutive turns in the plan (if the distance between them is less than 300 m).

Table 1.7.

Speed, km/h	30	40	50	60	80	100	120
The minimum visibility distance that ensures safe movement at a specified speed	80	100	120	150	200	280	350

Steep slope 1.13 and Steep slope 1.14 signs are installed in cases where the length of the slope is greater.

Table 1.8.

Slope, %0	40	50	60	70	80	80+
Slope length, m	600	450	350	300	270	250

Road narrowing signs 1.18.1-1.18.3 are installed in front of road sections where the roadway in populated areas is 0.5 m, or if it is reduced by one or two lanes in populated areas. Also in non-residential areas, in front of these signs, bridges, overpasses, overpasses, tunnels, and other artificial structures, if the width of the road section within these structures is equal to or less than the width of the road section in front of them. In , , and settlements, this indicator is set only if it is small.

In non-residential areas, all signs 5.16.1, 5.16.2 for pedestrian crossings are installed and (or) indicate unregulated pedestrian crossings in front of road markings 1.14, as opposed to crossings in populated areas if they are less visible. Above 150 m, Pedestrian crossing sign 1.20.

Pedestrian crossing sign 1.20 is not permitted to be installed in front of crossings before intersections.

Children's sign 1.21 is installed in front of the section of the road where exit from the territory of children's institutions (schools, kindergartens, etc.) is possible.

Exit signs are used to indicate turning for crossing intersections, crossing traffic lanes, or narrow road sections. The main road sign 2.1 is installed at the beginning of the road and before each intersection in populated areas.

Signs 2.3.1 and 2.3.2, 2.3.3 are used to give priority to drivers on roads marked with sign 2.1, who must cover a distance of 150–300 m before the crossing is installed.

Pass sign 2.4 is installed at the entrance to the main road if the intersection is clearly visible. These signs must be installed in previously uninhabited areas at a distance of 150–300 m from the intersection (unpaved road).

Prohibition marks are used to place or remove certain restrictions on movement.

The maximum speed limit sign 3.24 is installed depending on road conditions in the following locations:

1. In front of road sections where visibility for oncoming vehicles is not ensured (on curves in the longitudinal part of the plan), it must be established in accordance with Table 1.9;

Table 1.9.

Visibility of oncoming vehicles, m	Allowable speed, km/h
100 less	40
100 - 120	50
120 - 150	60
150 - 200	70
200 - 250	80

In settlements, according to Table 1.10, it is established in the following cases:

Table 1.10.

Distance from the edge of the roadway to the construction line, m	Permissible speed in a settlement, km/h	
	Less than 1 km	More than 1 km
Less than 5	50	40
5 - 10	60	50
10 - 15	60	60
15 - 25	70*.	60
25 - 35	80*.	70*.

At pedestrian crossings in the cases specified in Table 1.11;

Table 1.11

Pedestrian traffic, person/hour	Allowable speed, km/h
Less than 50	60
50 - 100	50
100 more	40

It is placed in front of small bridges under the following conditions.

Table 1.12.

The roadway on the bridge and the width of the roadbed	Allowable speed, km/h

Equal	70
1 meter smaller	50

It should be noted that the minimum maximum speed limit on highways must be at least 40 km/h.

Command signs define the movement of vehicles in a specific direction in an abstract order and mode. They are installed in front of places where it is necessary to change the direction and mode.

If it is necessary to use a prohibiting or commanding sign when organizing movement, it is always advisable to use a command sign.

No. 3.20 is placed depending on the intensity of traffic, its composition, road conditions, visibility, and weather conditions in the following cases.

Table 1.13.

Road to be installed signs	Width of the carriageway, m	Depending on the condition of the pavement, signs 3.20 or 3.22 are installed if the traffic intensity (m/h) exceeds the specified value.	
		dry	wet
Overtaking trucks is prohibited 3.22	6.0 - 6.5 7.0 - 7.5 9.0 - 9.5	300 500 700	170 300 400
No overtaking 3.20	6.0 - 6.5 7.0 - 7.5 9.0 - 9.5	500 750 800	300 500 700

Under certain conditions, it is difficult to solve the problem using imperative signs. For example, the use of sign 4.1.1 makes turning impossible, which creates certain difficulties. Therefore, it is correct to use the notation 3.18.1.

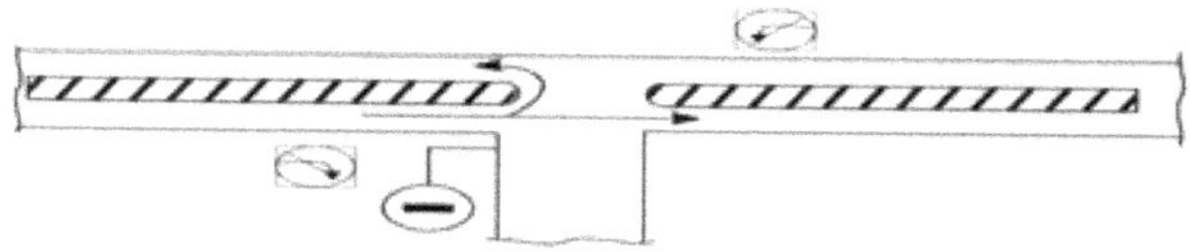

Figure 1.5. The use of a command character is inappropriate Condition

Information and direction signs are used to enter or exit certain traffic modes, as well as to provide information on the location of residential buildings and other objects.

Highway sign 5.1 is placed at the beginning of roads where the highway meets certain requirements (these requirements are specified in the highway course).

Within the boundaries of the existing building, the beginning of the residential zone is 5.22 and the end of the residential zone is 5.23 (with a white background). Under these conditions, the speed of vehicles leaving populated areas should not exceed 70 km/h.

On one-, two-, and three-lane roads, it is permitted to install sign 5.23 in both directions behind sign 5.22.

The beginning of Hal-Joy is 5.24 and the end of Fear is 5.25 (cleared background). The conditions for installing these signs are the same as for the previous ones. 5.24, the speed limit in populated areas is not limited to 70 km/h. Sign 5.24 indicates the type of settlement located off-road and is installed primarily in cases where residential buildings in the settlement are located 200 m or more from the road.

Service marks serve to provide information regarding the location of relevant objects.

If the distance between filling stations is less than 60–80 km, the first mark is not placed.

Service signs in residential buildings must be installed in advance at a distance of 100–150 m from the facility.

Additional information signs (plates) are mainly used to identify or limit the impact of the signs used with them.

The lines of road signs are divided into two groups: horizontal and vertical.

Horizontal road signs are used on improved roads and are divided into:

- longitudinal lines of the signs;
- transverse lines;
- safety island;
- inscriptions and other lines indicating the place of residence of the population, route direction.

Horizontal road signs are used on highways and highways in populated areas, as well as on roads used by public passenger transport.

1.2 Requirements for technical means

Road safety requirements for the technical condition of existing vehicles are provided in GOST 25478-91. According to statistical data, 40-45% of accidents occurring for technical reasons are caused by malfunctions in the vehicle's braking system. Traffic management plays an important role in ensuring smooth and efficient transport in cities and beyond. With the advancement of technology, technology has become an indispensable tool for optimizing transport flows, increasing safety, and improving overall mobility.

One of the main requirements for technical means of traffic control is the ability to monitor and collect data in real time. This includes information about vehicle volume, vehicle speeds, traffic levels, and traffic accidents. Accurate and timely data collection allows road managers to make informed decisions, optimize traffic light timing, and respond quickly to changing traffic conditions.

The implementation of intelligent traffic management systems is essential for efficient transport management. These systems utilize advanced algorithms and sensors to optimize traffic signal timing, prioritize specific traffic flows, and improve overall traffic flow. Smart traffic management systems help reduce congestion, reduce travel time, and improve road safety.

Adaptive traffic light control is another key requirement for vehicle controls. These systems use real-time data and predictive analysis to adjust signal timing based on current traffic conditions. Adaptive traffic light management can optimize traffic flow, reduce delays, and improve overall intersection efficiency, especially during peak hours.

Effective detection and management of road traffic accidents is essential for minimizing violations and ensuring the safety of road users. Technical tools such as traffic cameras, sensors, and event detection algorithms help quickly detect events such as accidents, road closures, or parked vehicles. Rapid response and effective event management are crucial for maintaining traffic flow and preventing duplication.

Integrating traffic management technology with public transport systems is crucial for developing multimodal transport and reducing dependence on personal vehicles. Continuous coordination between traffic lights, bus schedules, and other forms of public transport helps improve connectivity, reduce congestion, and create sustainable travel opportunities.

Monitoring and performance evaluation.

To evaluate the effectiveness of traffic management strategies, technical means must support monitoring and evaluation of performance. Data analysis tools, performance metrics, and modeling techniques help analyze traffic patterns, identify challenges, and assess the impact of various activities. Continuous monitoring and evaluation are necessary to improve traffic management strategies and increase the overall efficiency of the system.

Malfunctions in the braking system are primarily determined by diagnostics. The braking system diagnostics process is carried out based on operating conditions and a list of characteristic malfunctions, during which appropriate diagnostic parameters are selected, standard indicators are determined, and diagnostic technology is developed. Diagnostic parameters are determined by analyzing systemic-causal forms compiled on the basis of a list of brake system malfunctions.

Diagnostic and regulatory indicators must ensure the continuous operation of the braking system, a specific braking distance, and vehicle deceleration regardless of operating conditions. The following main operational requirements are imposed on the braking system of motor vehicles:

the shortest braking distance at a given speed;

simultaneous operation of brakes on all wheels;

short-term activation of the braking system;

equal braking forces on the left and right wheels.

Diagnostic parameters of the braking system can be divided into two types based on the occurrence of malfunctions and design parameter indicators: general and elementary diagnostic parameters. General diagnostic parameters include braking distance and vehicle deceleration, brake force differential, and wheel values. The main diagnostic parameters include axle pressure, increase or loss of braking force, brake activation time, brake chamber impact, free play of the axle, compressor efficiency, etc. Diagnostic parameters of the braking system are measured in the following cases[1]:

while the vehicle is in motion;

use diagnostic tools permanently installed in the car;

use of brake test devices in stationary conditions.

The main diagnostic parameters of a vehicle's performance relative to the braking system are:

t — braking distance, m;

Rt — braking forces, N;

Sc - deceleration path, m;

Tc — deceleration time, sec;

Jc - deceleration rate, m/s2.

One of the main activities in assessing the technical condition of motor vehicles is determining the technical condition of units and mechanisms that directly or indirectly affect their safety. The technical condition and performance of the vehicle's braking system are determined by road and workshop testing methods according to GOST 25478-91. Test it in practice. Road testing is primarily used to assess the overall quality of the braking system. Brake system diagnostics are performed on a flat, dry, horizontal, and stationary road. On a dry asphalt road with a adhesion coefficient of at least 0.6, the car's speed is increased to 40 km/h, then it is braked, and the braking distance and deceleration are determined. Slowing is measured using a slowing device. Table test. At motor transport enterprises and service points, diagnostics are primarily carried out on stands. The stand is constructed in such a way that it organizes movement on the road. Diagnostic technology: after installing the vehicle on the stand, the wheel speed is increased to 50-70 km/h, the brakes are pressed sharply, and all clutches on the stand are disconnected. In this case, there is a force of inertia between the wheels and the drums against the braking forces, and after some time, the drums and rollers stop rotating. The braking distance is measured by the number of revolutions of the drums or the duration of their revolutions, while deceleration is measured by an angular decelerometer. A decelerometer is a device for measuring deceleration. The operating principle of the instrument is based on recording the motion of the inertial mass moving within it relative to the body. This displacement occurs due to inertia and is proportional to the vehicle's deceleration. The tires and wheels have a big impact on the safety of the car. Tire pressure must not be lower or higher than standard values. The tread height of the remaining tires must be at least: 1.6 mm for passenger cars; 1.0 mm for trucks; and 2.0 mm for buses. It is prohibited to use tires that are torn, with torn cords and protectors separated into layers, or to place different types of tires on the same axle (axle) of the vehicle.

CHAPTER 2. IMPLEMENTATION OF TECHNICAL MEANS FOR MOVEMENT ORGANIZATION AND REGULATION

2.1 Investigation of road traffic organization

Road safety plays an important role in ensuring the safety of our roads. As the number of vehicles on the roads increases, proper planning and management are essential to prevent congestion, accidents, and delays.

The importance of traffic management:

Effective traffic management is important for several reasons. Firstly, it helps reduce traffic congestion, which not only saves passengers' time but also reduces fuel consumption and emissions. Secondly, proper road safety management improves road safety by minimizing the risk of traffic accidents and ensuring proper movement. It also improves the overall quality of life for drivers, making driving more pleasant and stress-free.

Effective traffic management strategies:

There are several strategies that can be used to improve traffic control on the roads. One of the main approaches is the implementation of intelligent transport systems (ITS), which use technology to monitor and manage transport flows. This includes traffic light optimization, real-time traffic tracking, and dynamic route management to guide movement through congested areas.

Another important strategy is the proper design and maintenance of road infrastructure. This includes building a well-planned road network, designing efficient intersections, and installing appropriate signs to guide drivers. It is also important to ensure compliance with traffic rules and regulations and to prevent reckless driving.

Public awareness and education also play an important role in promoting safe and responsible driving. Activities that emphasize the importance of complying with traffic rules, treating other road users with respect, and teaching defensive actions help foster a culture of road safety and reduce the likelihood of accidents. It should be noted that in our country, constant attention is paid to the problems of ensuring road safety. The work being carried out in this direction contributes to the annual reduction of road traffic accidents. In particular, during the years of independence, the laws and regulatory legal acts of the government related to the sphere have been updated in accordance with modern requirements, constantly updated and supplemented, which has a positive impact on solving the problem.

For example, while the Law of the Republic of Uzbekistan on Road Safety was adopted in 1999, it was revised and re-adopted in 2013 taking into account local conditions and, possibly, the suggestions and wishes of road users. In addition, it should be noted that 12 amendments and additions have been made to the articles of the Code of the Republic of Uzbekistan on Administrative Responsibility in the field of road traffic.

Today, as noted above, the reduction in the number of road traffic accidents, injuries, and fatalities is a positive result of efforts in this direction.

Undoubtedly, the role of the Traffic Rules in increasing the efficiency of road safety is invaluable. Therefore, the rules are constantly being improved, taking into account local conditions and accumulated experience in the field of road safety.

In recent years, the Traffic Rules have been amended and updated several times, taking into account local conditions. The main goal is to increase the efficiency of ensuring road safety.

The newly adopted Traffic Rules of 2015 were prepared for the same purpose. Its draft was prepared and studied by experienced responsible specialists from all official departments and organizations of the ministry.

Fifteen amendments have been made to the new edition of the Traffic Rules. In addition, a number of changes have been made. Certain terms and concepts have been clarified. Let's look at some of them to make their essence clearer.

If you have noticed that the word "auto train" is often mentioned in the previous and adopted Traffic Rules, that is, in their text, there is also a sign of its recognition. For this reason, the term "road train" was included in the General Part of the National Railways (Section 1) and its concept was clarified. Delivery package includes tractor and semi-trailer or trailer (s). This term was introduced taking into account the movement of such trains across the territory of our country.

In the Road Traffic Regulations currently in force in the Republic of Uzbekistan, the term "motorcycle" is defined as a two-wheeled mechanical vehicle with or without a saddle. They explained that three-wheeled motorcycles with a gross weight of no more than 400 kilograms are also considered motorcycles.This explanation could not fully satisfy the existing demand.Because today, as a result of technological progress, there are cases of traffic violations on the roads of two-, three- and four-wheeled motor vehicles that are not called motorcycles, but belong to the same category, that is, modern coupe cars, scooters and electric motors.To regulate traffic rules, it became necessary to make appropriate changes to this rule.

Therefore, the concept of a motorcycle was expanded to include scooters, electric scooters, and ATVs with an engine displacement of more than 50 centimeters and speeds exceeding 50 kilometers per hour. Registered vehicles will now be registered as motorcycles and assigned license plates. Drivers operating this type of vehicle must hold a driver's license granting the right to drive Category A vehicles.

- So, until now, it was possible to drive such cars without license plates and driver's licenses, right?

- Yes, it's impossible now. In the event of such an offense, drivers are subject to a fine in the amount of five times the minimum monthly wage introduced by the Administrative Responsibility Code of the Republic of

Uzbekistan; in the absence of a driver's license, a fine in the amount of three times the minimum monthly wage introduced in accordance with Part 2 of Article 135 of the Code is applied.

As I mentioned above, some phrases in the previous Traffic Rules could not find their full expression. One of them is the phrase "yol marzasi," meaning the edge of the road. This phrase is often found in the Traffic Regulations. Therefore, everyone who read it understood it in their own way. This shortcoming has been eliminated in the new rule. The road boundary is located at the same level as the traffic line and is separated from it by line 1.1 or another surface designated as an element suitable (in accordance with the requirements of these rules) for driving, stopping, or parking vehicles.

It is well known that ensuring the safety of passengers, especially children, transported by vehicles is an extremely responsible matter. For this reason, the term "organized transportation of a group of children" was introduced into the Traffic Rules. (except for targeted vehicles) is defined as organized transportation of children. We believe that the implementation of this concept will place a great responsibility on the heads of schools and preschool educational institutions, colleges, and orphanages, including representatives of our transport service.

Those who have read the traffic regulations carefully know that the word "real weight" is frequently encountered in them. This word is mainly derived from transport regulations and state standards. Therefore, the term Actual weight is included in the Rules to determine the actual weight of the vehicle (in fact). Thus, by actual weight, we mean the actual weight of the vehicle along with its cargo, driver, and passenger. Determining road routes based on road capacity is of great importance.

Special attention is paid to the introduction of information and communication technologies. It is well known that in the next 5 years, the practice of automatic detection of traffic violations will be widely used in the activities of the road patrol service. Therefore, the term "photo and video recording" is reflected in traffic regulations and is also recorded in administrative documents.

Additionally, a Photo and Video Information Sign has been introduced to warn drivers. The relevance of the introduced innovation is closely linked to the launch of a computerized road safety management system in Tashkent.

Now this innovation is working in the online system. We fight offenses not by punishing road users, but by warning them and calling them to follow the rules.

The new rules impose a number of obligations on drivers; what are they?

It is true that the new edition of the Traffic Rules imposes a number of important obligations on drivers. Now, before driving, the driver must warn passengers not to remove their bodies from the passenger cabin while driving (this does not include putting their hands out). The same requirement has been added to paragraph 24 of the passenger regulations. You know, lately, you can see people taking nude photos, especially at wedding ceremonies. This has a serious impact on traffic safety. That's why this change was made. Furthermore, the driver must notify passengers of the need to fasten their seatbelts.

At this point, it would be appropriate to mention some considerations. These changes are not taken for granted. They are carried out as a result of many years of research, research, analysis and study of the opinion of the participants of the movement. Therefore, this news should be studied not only by industry workers but also by the general public, parents, and teachers of schools and colleges. Our people did not say for nothing that traffic rules are a blessing in life. It will be useful if it is studied carefully and strictly observed. If a parent, school or college teacher, or traffic police officer calls a pedestrian to clean the house, a positive result will certainly be achieved. I want everyone to pay attention to these aspects of the issue.

Speed is another factor that has received special attention in the new version of the Traffic Rules. You know, the number of vehicles is increasing day by day. The increase in the number of vehicles leads to traffic jams. It is well known that violating speed limits can lead to road traffic accidents with severe consequences. Based on the analysis, increasing the speed in certain areas can save fuel consumption, travel time, and prevent environmental pollution. Taking into account these aspects of the issue and the opinions of relevant specialists, traffic in residential areas has been regulated, and in order to ensure public safety, a high speed of 30 kilometers per hour has been established between apartment buildings. Furthermore, by the same clause, in coordination with the Council of Ministers of the Republic of Karakalpakstan, the khokimiyats of the regions and the city of Tashkent, and the State Road Safety Service, it is permitted to increase the speed limit for certain vehicles if the condition of the road allows for increased speed in certain places or on certain sections of the road.

- In many countries, cars turn on their headlights even during the day. This must have been done for security reasons, right?

- You're right. Vehicles attract the attention of many people when they turn on their headlights even during the day. This is also taken into account in the new rules. Now, in order to strengthen safety measures, buses, minibuses, as well as motorcycles and mopeds carrying people and transporting passengers, are required to turn on their headlights even when moving during the day, and this procedure is based on paragraph 138 of the rules.

Furthermore, Article 149 of the Rules, which establishes the age of vehicle drivers, has been amended in accordance with the requirements of the Regulation on Driver Training. Therefore, according to the rules, the age of those studying to obtain a driver's license for the right to drive Category A vehicles upon graduation is 16 years, and the age of those studying is 16 years. Obtaining category B and C driver's licenses can be from the age of 18 at the time of graduation, D and those who are studying to obtain a driver's license giving the right to drive a tram or trolleybus. Must be 21 years of age. Graduation years.

2.2 Research on the installation of road signs and road markings on roads

Road users are required to know and comply with the Law of the Republic of Uzbekistan "On Road Safety," these Rules, and the requirements for traffic lights, road signs, and road signs contained therein, as well as to comply with the established norms. This was reported by the Road Safety Department. In recent years, special attention has been paid in our country to eliminating existing problems in ensuring road safety, improving road transport infrastructure based on international standards, reconstructing highways, and increasing the convenience and safety of road traffic.

In a number of decrees and resolutions of the head of our state, new tasks and measures for the comprehensive improvement of road infrastructure have been defined. This gave impetus to the further development of the industry.

Including:

1) ensure the implementation of the Decree of the President of the Republic of Uzbekistan dated July 12, 2022, No. 5. a number of measures have been implemented.

In particular, for the period ending in 2023:

15,621 new road signs were installed and 35,984 unusable road signs were replaced;

6,531 speed limit signs and 2,959 direction and distance signs have been installed in front of schools;

157 new traffic light facilities were built and 85 were reconstructed;

78 new pedestrian traffic lights were built, 17 were reconstructed;

10 square meters of road lines were built. km;

Artificial road lighting was installed in 269 km of settlements, and pedestrian and bicycle paths were built in 301 km of settlements;

means (barriers) restricting the movement of pedestrians are installed in places where crossings are not established, at a distance of 91 kilometers from settlements;

245 km of dividing barriers (BDO-5) and iron barriers were installed in opposite directions.

10 km2 (plan 12 sq. km) of road sections were repaired and 35 (plan 45) intersections

were landscaped by changing their geometric parameters;

25 turns on the roads (29 planned) were replaced with drip turns;

Eight checkpoints have been installed to monitor the weight and volume of vehicles.

Article 128 of the Code of Administrative Responsibility of the Republic of Uzbekistan (Drivers operating vehicles on pedestrian paths who do not comply with the requirements of road signs and other traffic rules), drivers of vehicles moving on pedestrian paths who do not comply with the established norms. Violation of requirements for road signs or markings on the roadway (except for cases provided for by Article 1283, part one of Article 1284, Articles 1285, 1286, and 130 of this Code), the transportation of people, the overtaking of vehicles, the bypass of stopping places where vehicles pass, the use of parking or pedestrian crossings, external lighting devices, as well as the splashing of passengers with mud — shall entail the imposition of a fine in the amount of two base calculating values.

Committing an offense provided for in part one of this article repeatedly within a year after the application of an administrative penalty shall entail the imposition of a fine in the amount of one base calculation unit.

Committing the same offense provided for in the first part of this article twice within a year after the application of an administrative penalty shall entail the imposition of a fine in the amount of three times the basic calculated amount.

Committing the same offense provided for in the first part of this article repeatedly three or more times within a year after the application of an administrative penalty shall entail the imposition of a fine in the amount of four times the basic calculated value or the deprivation of the right to drive a vehicle for six months.

Lines indicating the state of the road at the edges and elements of the roadway on the roadway (with solid pavement) may be used together with road signs and traffic lights or separately.

Road markings are widely used in traffic management and are drawn where the driver's primary attention is focused. Road lines are divided into groups of horizontal and vertical lines. Each group is designated by its own number. The first digit indicates the group (the 1st horizontal and 2nd vertical lines), the second digit the order of the road signs in the group, and the third digit the appearance of the road signs.

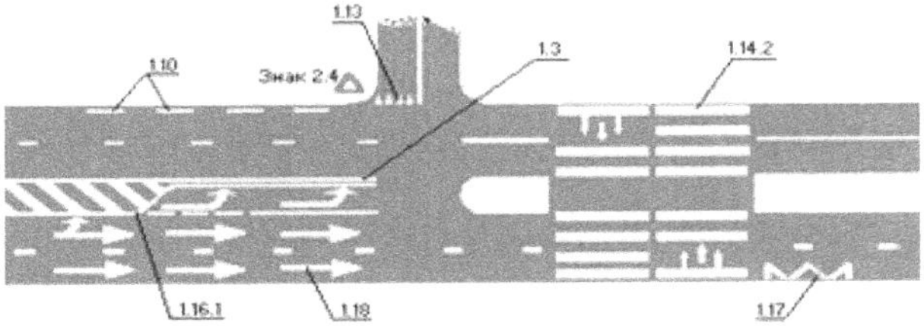

Figure 2.1. Horizontal and vertical lines.

It is known that the lines of road signs are divided into horizontal and vertical. Horizontal lines primarily determine the order of movement. The horizontal stripes are white and yellow. The white horizontal stripes themselves are solid, ring-shaped inscriptions and numbers that differ from one another in size. There are three types of yellow horizontal lines: continuous, circular, and zigzag. It is necessary to know the places where they are drawn and their meaning.

A narrow white line drawn in the middle of the road separates the oncoming traffic. Crossing the line is prohibited.

A narrow, smooth white line in the middle of the road separates the oncoming traffic. Crossing the line is prohibited.

Figure 2.2.

A continuous white solid line drawn between sections of the roadway in hazardous areas marks the boundaries of the sections.

Crossing the line is prohibited.

On highways, a wide (20 cm) continuous solid line drawn along the edge of the roadway marks the boundary of the roadway.

The line can be crossed from any side.

Figure 2.3

Two parallel solid white lines in the middle of the road separate the flow of vehicles in opposite directions on roads with four or more sections. Crossing the line is prohibited.

Figure 2.4

A yellow line drawn along the edge of the roadway or on a curb (bordyur) prohibits the stopping of vehicles. It does not apply to fixed-route transport.

The line can be crossed from any side.

Figure 2.5.

A narrow dashed white line drawn in the middle of the roadway marks the boundary of the sections on roads with two or more sections (the ratio of the distance to the line is 1/3). The line can be crossed from any side.

Figure 2.6-2.7.

A dashed line drawn next to a solid line is permitted to be rearranged only in one part.

Crossing this line:

a) possible from the discontinuous side.

b) is prohibited from the side of the solid line.

Figure 2.8.

Determines the boundary of the carriageway in reverse traffic.

On roads where reverse traffic is carried out (when the reverse traffic light is turned off), it separates the traffic flow in the opposite direction.

In the absence of reverse traffic lights or when they are switched off, crossing line 1.9 is permitted only if it is located to the right of the driver.

Figure 2.9.

When a reverse traffic light is turned off, it is prohibited to intersect lane 1.9 in the opposite direction with a split traffic flow.

Vertical lines consist of a complex of linear, white, and black lines applied sequentially along the edges of road structures and equipment, serving as a means of visually determining their dimensions.

The sky is cloudy, gray, and dark; on unlit sections of the road, the color of the vertical concrete structures on the roadside closely resembles the color of the sky and the road, and the driver cannot distinguish them without seeing the remains of the concrete structures while driving. This situation leads to the driver hitting concrete structures.

Road markings are applied to the surfaces of bridges and road columns, allowing the driver of a moving vehicle to clearly distinguish the color of a hazardous obstacle on the road and make it easier to see.

Road marking lines are drawn on the surfaces of bridges and suspension columns to sharply highlight the color of the hazardous obstacle on the road and facilitate visibility for the driver of the moving vehicle.

Figure 2.10.

Black-and-white drawings of road markings on the lower edges of tunnels, bridges, overpasses, and bridges make it easier for the driver to distinguish the lower part of the structure.

Figure 2.11.

Road markings refer to circular posts on separating sections or safety islands.

Figure 2.12

The road sign indicates concrete or reinforced concrete guide posts, barrier supports, and piles, clearly showing the trajectory of the road with its sharply contrasting black-and-white color when it is dark, in thick fog, and when everything is covered with white snow.

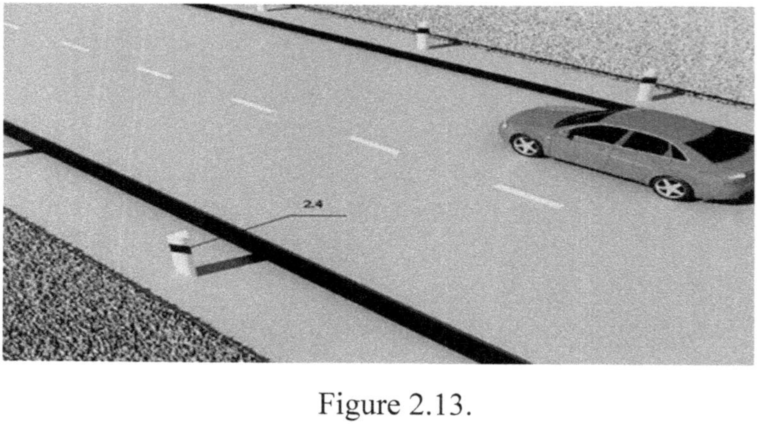

Figure 2.13.

The lines of road signs represent a small radius and sharp turns in the road.

installed on roadsides, slopes, and other dangerous places

shows the sides of the obstacles. Road marking lines indicate the curb in hazardous areas and safety islands above the road surface.

Horizontal lines

Lines, signs, inscriptions, and other signs on the road surface determine the order and precise order of movement.

The lines can be permanent or temporary.

If it is impossible to see the road markings (snow, mud, etc.), road signs corresponding to the meaning of these markings must be installed on this section of the road.

The horizontal lines will be white (lines 1.4, 1.10 and 1.17 are yellow), as well as line 1.24 (2x2 meters with a yellow image of a person with a disability on a blue background).

Pedestrian crossings may be white or other colors in combination with white.

CHAPTER 3. DEVELOPMENT OF RECOMMENDATIONS FOR THE APPLICATION OF TECHNICAL MEANS FOR THE ORGANIZATION AND REGULATION OF ROAD TRAFFIC

3.1 Ways to install modern technical means for traffic management

Traffic management plays a decisive role in solving many social problems. This has proven effective in ensuring safer driving, reducing pollution, increasing accessibility, and helping governments upgrade existing infrastructure. Most traffic controls are things we see every day, such as traffic lights and right turn routes, while others are less visible and more complex. By the 1930s, America had already become an automobile society, and the growth of urban transportation played a major role in the development and approval of the interstate highway system. It soon became clear that standard planning was not enough to solve the problem, and authorities began looking for technologies that would make the roads more efficient.

Although traffic management solutions are almost as old as the car itself, the history of traffic management technology as we know it today actually dates back to the 1960s.

Highway departments began using ramp counters to limit the number of interstate vehicles at a time. Speedometers, like many later controllers, are effective in some respects and limited in others.

However, this creation served as a catalyst for large and effective developments in road technology for many years.

decades of research have shown the value and necessity of traffic management solutions.

The further development of these technologies plays an important role in solving the problems of current and future transport systems.

If we take a look at the technologies available, it shows that we have potential for our future.

In fact, research by the Organisation for Economic Co-operation and Development on the impact of road safety technologies suggests that future changes could reduce deaths and injuries by 40%.

Traffic management technologies can benefit all passengers, from drivers to cyclists and public transport users, by creating safe, smart, and efficient roads.

These benefits extend beyond transportation and have been proven to have positive economic and environmental impacts.

Let us consider the main advantages of intelligent traffic management technologies.

Reduce data and save time

By 2025, the average American passenger is expected to spend 62 hours per year in low-traffic conditions [7].

This means losing more than two and a half days of time.

Traffic management technologies create more efficient systems that reduce both the number of drivers on the road and the time each person spends driving or in traffic.

This is beneficial for both the passengers and the Earth.

From roadside cameras to 511 information systems to mobile phone applications, the same ITS technologies have been proven to help significantly reduce pollution.

Cities using these technologies have reduced fossil fuel consumption by 53 million gallons and CO2 emissions by 10 billion pounds per year.

Road technology also collects data on street usage and driving patterns that will influence future infrastructure development.

This will help planners create safer street systems for drivers, pedestrians, and cyclists.

Expanding these technologies is crucial for reducing bicycle crashes and pedestrian deaths and developing street plans that support multiple modes of transport.

The use and expansion of traffic management technology will help cities and states create better economies.

Traffic management technologies have been proven to reduce congestion and help solve these problems [11].

Traffic management technologies also facilitate the development of walkable cities, where high real estate values, high government spending rates, affordable vehicles, low healthcare costs, and a talented workforce play an important role.

Planners and policymakers are introducing new, innovative transportation management systems to reduce congestion and create safer, more successful communities. These innovative solutions are applied alongside traditional methods and provide city officials with new ideas for making roads more efficient and sustainable.

These sensors are installed on roads and public transport routes, allowing anyone with internet access to see traffic levels, available stops, and the location of buses and trams.

Cities are equipping government vehicles with V2I devices that transmit speed, individual vehicle traffic, time spent at traffic lights, and other valuable information.

This information provides officials with information on how long cars are idle at empty traffic lights, so they can improve signal timing to make driving time more consistent and efficient.

These vehicles and sensors transmit information about road conditions to drivers via radio, cellular communication, and fiber optics to improve road conditions.

Advancements in artificial intelligence and its application are driving forces in almost all aspects of modern life, and in many cases, they can even be called game-changers. This is especially true for mobility, where artificial intelligence allows for the creation of new applications that support and activate traffic flows.

A tool that enables the use of artificial intelligence in road applications is easy access to more powerful hardware. This includes, on the one hand, faster, smaller, and more efficient processors that allow for more operations in an increasingly smaller area, and on the other hand, advanced camera technology, which has made great strides in recent years.

Today's industrial cameras have more capabilities than the human eye—from color detection and a wide field of view to seeing very fine details and in the infrared spectrum. In addition, they allow you to take excellent photos even in conditions of poor visibility, for example, at night or in bad weather. The first thing you get with this tool is the software that determines the possibilities of using video traffic. It decides what content can be processed, analyzed, and used, in other words, how to present the ideal presentation to applications. Smart decisions are important. Neural networks such as CNN (Convolutional Neural Networks) used in Vitronic are particularly promising approaches in this regard. These networks can be trained using images and provide high-performance detection and identification algorithms. Literally in milliseconds, they can achieve more differentiated or more efficient objects. It is necessary to consider two areas that demonstrate the advantages of artificial intelligence for road traffic. The first is road safety—more precisely, the prevention of accidents. For example, intersections and intersections already account for nearly half of the injuries and deaths of urban road users. As a result of the transition to road traffic, especially in these transport hubs, traffic scenarios are becoming more complex, and more and more different road users, including micromobility, are interacting with each other. This further increases the risk of accidents.

AI photo radars

Figure 3.1.

Video-based law enforcement systems using artificial intelligence are well-suited for these changes, as they can monitor more complex situations involving different road users. For example, they can automatically distinguish between pedestrians, cyclists, and cars, thereby detecting line violations at pedestrian crossings. This option significantly improves the safety of vulnerable road users (VRUs). Furthermore, various disorders can be detected simultaneously, thereby resolving the causes of several accidents simultaneously. Possible traffic control applications include pedestrian crossing control, red light control, turn warning, and more. One of the innovations in this field is a fully automated execution system using artificial intelligence software that detects smartphone usage while driving. This is the answer to a very modern problem, because these behaviors are common and extremely dangerous. Recent studies show that the risk of accidents increases twofold when talking on the phone, sixfold when sending texts, and twelvefold when dialing a phone number. Many road users underestimate this risk, and it did not have much deterrent effect, as enforcement requires too much manpower and is almost impossible to implement across the entire council.

The new system changes this with its mobile as well as semi-desktop apps and offline features. In this application, the artificial intelligence is trained using images so that it can detect whether the vehicle driver is holding a mobile phone, replacing the time-consuming manual assessment of image recordings. Thus, tracking a mobile phone distraction is now possible at any time, anywhere, and can change a driver's behavior in the medium and long term.

The AI-based system has another application that contributes to road safety. It can be used to detect breaches of seat belts. Although this does not prevent accidents, it can have a significant impact on the severity of their consequences. Wearing a seat belt reduces the risk of fatal injuries by at least 50 percent, and one in four drivers or passengers who died in a traffic accident was not wearing a seat belt or wore it incorrectly. Soon, this AI-based system or similar systems could be used to detect careless driving violations, thereby further improving safety.

Manage enhanced data

The second area that highlights the progress that AI will bring is traffic optimization in the urban environment. There is a problem of improving transport management in cities, making it more stable and reducing it. The figures below show how necessary this is. In 2022, drivers in major German cities were stuck in traffic for an average of 40 hours. In 2020,

96% of the EU urban population was exposed to air pollution that exceeded the WHO limit.

One way to change the situation is through traffic regulations and restrictions. This includes, for example, low-emission zones, public transport corridors, or vehicles where many people live. In this case, enforcement of the law often becomes a problem: how can we ensure that drivers do not use the bus lane? How can we ensure that only authorized vehicles enter the ecological zone? Or even more difficult: how do you know that drivers aren't driving alone in a crowded area? Once again, AI provides the solution. It can not only read license plates but also collect other information, ranging from the type of vehicle to the number of passengers. This will create a database that will help you take effective action. Given all of this, it is clear why artificial intelligence and video-based transport systems will help traffic flow. But the diversity of their application and operation is not the only reason. Two more should be highlighted. First, their innovative potential. For developers, this is an affordable technology with fewer barriers to entry than others, making it easier to creatively solve problems and further develop. Second, they require relatively low investment for acquisition and implementation, resulting in a faster return on investment. This is especially interesting for smaller municipalities that previously faced difficulties with refinancing. With the new availability of efficient transport systems, more states and municipalities can use them and in turn support more inclusive transport.

3.2 Proposals for improving road signs and markings in traffic management

Road sign coatings play an important role in improving road visibility, especially during inclement weather or at night. The reflective or luminescent materials used in these coatings enhance visibility and make road signs stand out even in poor lighting conditions. This expanded view is necessary to guide drivers and warn them of approaching hazards, such as pedestrian crossings, roundabouts, or merged lanes.

Figure 3.2. Location of road signs outside of populated areas.

Further improvement of the installation of road signs and markings outside settlements:

The paints used to ensure the good visibility of road signs and road markings must be of the highest quality. This will significantly improve visibility and enhance driver comfort. Visibility deteriorates significantly, especially at night outside of populated areas. The quality of the paint lasts a long time.

Along with the road sign, it is necessary to use improved instruments, that is, the use of electronic lights along with the sign makes it very convenient for the driver.

It is also proposed to reduce the brightness and color of advertising monitors installed at intersections. This creates inconvenience for the driver.

Coatings reflecting road signs not only benefit drivers but also make a significant contribution to the safety of pedestrians and cyclists. Pedestrian crossings, zebra crossings, and bicycle paths are marked with durable and highly visible paths, providing appropriate supervision and protection for vulnerable road users.

Properly designated pedestrian crossings help pedestrians move safely on busy roads, reduce the risk of accidents, and contribute to a pedestrian-friendly urban environment. Bicycle lanes marked with unique pedestrian paths separate cyclists from motor traffic, enhance their safety, and encourage more people to use bicycles as an eco-friendly mode of transport.

Effective road sign coatings must possess high strength and durability, especially considering the constant impact of transport, severe weather conditions, and regular maintenance such as street sweeping and snow removal. With proper application and care, high-quality coatings can withstand heavy traffic, resist drop-off, and remain visible for a long time.

Long-lasting road sign coatings not only reduce the need for frequent painting but also help save costs for highway authorities and municipalities. By ensuring that road signs are clear and understandable over time, these surfaces play an important role in maintaining a safe and orderly road network.

The covering of road signs is an indispensable tool for improving road safety and efficiency. They provide clear visual guidance to drivers, improve road visibility, and create a safer environment for all road users. By investing in high-quality, durable road signs, road authorities and municipalities can make a significant contribution to reducing accidents, improving traffic flow, and creating a safer and more user-friendly transport infrastructure.

As technology and materials continue to evolve, it's important to stay up to date on the latest advancements in gesture overlays. Ongoing research and development in this area will help find more efficient and sustainable solutions to meet the evolving needs of our constantly expanding road networks, ultimately making our roads safer for everyone [4].

CONCLUSION

The use of technical means of traffic management and regulation is of great importance in solving problems related to traffic congestion and road safety. By utilizing advanced technologies such as ITS, traffic light control systems, automated payment collection systems, and intelligent traffic management platforms, we can create a safer and more efficient road network for all road users.

1. The authorities should continue to invest in these technical solutions to ensure sustainable and efficient traffic management practices in the future. In today's rapidly developing world, transportation has become an important issue due to the increasing number of vehicles on the roads. Due to population growth and urbanization, the need to take effective measures for the organization and regulation of transport

traffic has become more urgent than ever. One of the most effective ways to solve this problem is the use of technical tools.

2. Technological progress has fundamentally changed our approach to traffic management. Various technical solutions have been developed to improve road safety and efficiency: from smart traffic lights to automated payment collection systems. These technologies not only help reduce traffic volumes but also play a crucial role in preventing accidents and maintaining traffic flow.

3. One of the primary technical means for organizing and regulating traffic is the implementation of intelligent transport systems (ITS). ITS uses sensors, cameras, and communication technologies to monitor and manage traffic flows in real time. By collecting and analyzing data, the ITS can provide traffic authorities with valuable information, allowing

them to make informed decisions to optimize traffic flows and reduce congestion.

4. Another important technical tool is the use of traffic light control systems. These systems are designed to adjust signal times according to transport conditions, helping to minimize delays and improve overall traffic flow. Adaptive traffic light control systems are capable of providing efficient vehicle movement through intersections and responding dynamically to changes in traffic patterns.

5. Having discussed international experience in road safety management, it should be concluded that practical measures to ensure road safety in foreign countries are primarily aimed at the material liability of offenders.

6. Further improvement of road signs and markings will be developed to create convenience for drivers when organizing traffic on the roads. For this purpose, it is necessary to

use road signs with electronic lighting or interchangeable signs, and it is recommended to paint the road markings with high-quality paints.

7. In the event of an accident, before providing first aid, you must pay attention to the safety of the victim. If bleeding occurs after the wound, it should be bandaged at a height of 15 cm from the bleeding site.

USED LITERATURE

1. K. Kh. Azizov. Fundamentals of safe movement on roads. Tashkent – 2005. 45-46.
2. Traffic rules. Tashkent - 2022
3. Road signs and road markings. Tashkent - 2023
4. K. Kh. Azizov. Fundamentals of safe driving on roads. Tashkent - 2005.
5. N. F. Mukhitdinov, G. K. Nurmukhamedov, R. N. Dimitov. Commentaries on Traffic Rules. Tashkent - 2005.
6. Traffic rules. Tashkent - 2001.
7. Road signs and road markings. Tashkent - 2001.

   8. K. Kh. Azizov. Fundamentals of safe movement on roads. Tashkent - 2005.
8. N. F. Mukhitdinov, G. K. Nurmukhamedov, R. N. Dimitov. Commentaries on Traffic Rules. Tashkent - 2005.
9. Khodjaev B.A. Fundamentals of Cargo and Passenger Transportation by Road.

10. Azizov K.Kh., Dorobov M., Sayfutdinova R.A. Ensuring traffic safety on city streets. - T; TADI. 2009. p. 132.
11. Babkov V.F. Road conditions and traffic safety; textbook for higher education institutions. – M; Transport, 1993. - 271 p.
12. www.gai.auto.ru.
13. www. exha.ru

CONTENTS

www.ingramcontent.com/pod-product-compliance
Lightning Source LLC
Chambersburg PA
CBHW050531240926
51846CB00003B/769
* 9 7 8 9 3 6 9 0 8 0 2 3 6 *